John Hunter, Gumperz Levison

An Essay on the Blood

In which, the objections to Mr. Hunter's opinion concerning the blood, are

examined and removed

John Hunter, Gumperz Levison

An Essay on the Blood
In which, the objections to Mr. Hunter's opinion concerning the blood, are examined and removed

ISBN/EAN: 9783337393205

Printed in Europe, USA, Canada, Australia, Japan

Cover: Foto ©berggeist007 / pixelio.de

More available books at **www.hansebooks.com**

A N

E S S A Y

ON THE

B L O O D;

IN WHICH,

The Objections to Mr. HUNTER's
Opinion concerning the Blood,
are examined and removed.

אך בשר בנפשו דמו׃

Verum tamen Carnem cum Sanguine ipfius, qui Anima eft
ipfius. *Gen.* ix. 4.

כי נפש הבשר בדם היא׃

Nam Anima cujuíque Carnis in Sanguine eft. *Lev.* xvii. 11.

כי נפש כל בשר דמו בנפשו הוא׃

Eft enim Anima cujufque Carnis Sanguis ejus pro Anima ejus
eft. *Lev.* xvii. 14.

By G. LEVISON, M. D.

L O N D O N:

Printed for T. DAVIES, in *Ruſſel-Street, Covent-Garden.*
Dead; ——————— Dead!
M. DCC. LXXVI.

T O

WILLIAM HUNTER, M. D.

Physician extraordinary to her

MAJESTY,

Professor of Anatomy to the ROYAL
ACADEMY,

Fellow of the Royal and Antiquarian
Societies, &c. &c.

SIR,

NEITHER partiality, nor
the expectation of your
patronage, as a supporter
of your brother's opinion, hath in-
duced me to offer these sheets to

A 2 your

your protection, being convinced,
that not only thofe of your brother,
but even your own opinions, are
held in no efteem with you, otherwife than they are conformable to
truth; and that you would not
hefitate to reject them the moment
their fallacioufnefs be proved; a
liberality and ingenuity to be found
only amongft *true philofophers,* and
which you candidly endeavour to
imprefs on the minds of your pupils,
always defiring them in your lectures, never to truft to any of your
notions, without the ftricteft ex-
amination, and declaring that you
can, with equal pleafure, renounce
any miftaken doctrine, as maintain
the true.

Duty

Duty and gratitude are the only motives, which move me to embrace this favourable occasion of acknowledging to the world, how much I owe both to your knowledge and friendſhip. For me to expatiate on your merits, would only be to leſſen them; the name of Dr. HUNTER ſtands in no need of commendation from my pen; only give me leave, Sir! to aſſure you, that I ſhall always think myſelf happy in finding every opportunity of convincing you, with how much pleaſure, and with what profound reſpect and eſteem, I am,

SIR,

Your very obedient,

And moſt humble Servant,

G. LEVISON.

THE

PREFACE.

S Phyſic is the moſt noble and important ſcience, ſo ought the terms in which it is couched to be the moſt perſpicuous and deciſive. It would indeed be a happy circumſtance, were the profeſſors of phyſic in general, to agree in uſing univerſal terms; many

diſputs

difputes have arifen, not only in the fchools
of phyfic, but in almoft all fpeculative
fciences, about mere terms only, which,
as foon as the ideas themfelves intended
by the words, have been candidly com-
pared, and the terms defined, opponents
agree, and the controverfy vanifhes immediately.

As long as we remain men, that is, as
long as we are not able to difplay and
unfold the true and real nature of things,
we fhall always labour under the burden
of controverfy in all fpeculative fciences;
for the fame effects will arife from caufes
feemingly different; and fince we reafon
of things by their properties only, with
which each obferver will be differently
ftruck, fome are apt to take for the effects,
what others confider as the caufes.

The word *life* has often put a ftop
to all reafoning at once, in both fciences,
phyfiology and pathology; all phænomena

mena are explained rather by a mere word *life*, than by a chymical or mechanical reafon: the different fecretions are faid to be owing to the different powers of life which each gland poffeffes, and that ftructure is entirely out of the queftion, where life refides; though it is very furprifing, that the wife creator fhould have formed each gland differently, when he could make them all alike, and only infufe in each a different kind of life; but may it not be afked, why is the power of generation loft, when there is a defect in either the male or female organs of generation? why did he not make the eye hear, the ear fee,——— were not the different ftructure of the organs themfelves neceffary to perform their different functions, which is mechanical, and were not different attractions of the juices in the different glands, neceffary to produce their various properties, which is chymical? life is certainly that quality, by which the very ftructure itfelf

is

is formed, by which it is set at work and
put into motion, which repairs and sustains
it continually; so that as soon as life is
gone, the structure and mechanism itself
is destroyed : but this life does nothing of
itself, without its necessary tool, mechanism: it is true, that no mechanism is
ever set in motion, without some cause
of life or its operation; but it is also undeniable, that life (as far as we know)
never operates without mechanism. And
is it not more confistent with reason to
suppose one power of life diffused through
the whole body, which operates differently
in the different parts, according to their
structure, than to attribute a different kind
of life to each part? We might as well
say, that each motion, in a different direction of the same muscle, is produced
by a different kind of life; both carry the
greatest absurdity in their face.

It is surprising, that we always are apt
to embrace extremes; thus lately every
thing

thing in phyfic was accounted for merely
by mechanics, whereas now, according
to fome, it is fufficient to know that
certain alterations may happen in the
body, that certain applications will remove certain difeafes, &c. without even
attempting to affign any reafon for the
different phænomena, forgetting the rule
of CELSUS, " *rationalem quidem puto me-*
" *dicinam effe debere : . inftrui vero ab evi-*
" *dentibus caufis; obfcuris omnibus, non a*
" *cogitatione artificis, fed ab ipfa arte re-*
" *jectis.*"

How far the different theories influence a different practice, I fhall illuftrate
by one inftance only : when mechanical
reafoning, and the applications of mechanical laws, were the fole means of ex-
plaining all alterations in the animal body,
it was certain, according to mechanical
laws, fince all the veffels of the animal
are fo many tubes which communicate
with one another, that when the fecretion

tion of one fet of veffels was increafed, the fecretion of another muft be diminifhed ; a falivation confequently was ftopped by either the exhibition of cathartics, fudorifics, &c. whereas according to fome moderns, the falivary glands have fomething peculiar to themfelves, to increafe their fecretion by the exhibition of mercury, which cannot be ftopped by the increafed fecretion of any of the other veffels.

The doctrine that the blood is alive, though very ancient, (a) and largely
treated

(a) It is repeatedly explained, Gen. ix. 4. and Lev. xvii. 11. 2. and 14. which I have ufed as my motto, that the נפש nephefh, which fignifies motion and growth, or increafe, and which is tranflated life, has its feat in the blood דם dam, fignifying affimulation, not in the בשר bafar, flefh or fibres; which cannot be confidered as accidental, as will appear from the following demonftration.

The

treated of by the learned Dr. WILSON in his lecture on the circulation, yet has never

---

The Hebrew has three words expreſſive of life, namely, נפש *nepheſh*, רוח *ruach* and נשם *neſhem*, which I imagine has given riſe to the doctrine of the ancients, concerning three different ſouls; each of theſe three words however denotes a different property, reſulting from the principle life; the firſt נפש *nepheſh* expreſſes growth and motion, the ſecond, רוח *ruach*, breath, and the third, נשם *neſhem*, breathing: the term חיים *chaim*, which is alſo tranſlated life, is quite different from the three fore-mentioned, in its meaning and uſage, ſtanding always by itſelf, and never uſed in regimen of the others, whereas each of them is often placed in conſtruction with חיים *chaim*, which proves that by חיים *chaim*, is not meant the progreſſion or reſult of life, but the principle, life itſelf, which principle is diffuſed through all nature, called by the ancients the *anima mundi*, and which is the foundation of the other three, expreſſing the effects and progreſſions, rather than the principle itſelf: whenever life is attributed to the blood, neither רוח *ruach* nor נשם *neſhem* is joined with דם *dam*, but always נפש *nepheſh*; becauſe the two former רוח *ruach* and נשם *neſhem*, which expreſs breathing and breath, are progreſſions of life, not exiſting in the blood, but on the contrary, give the blood

its

never been so much extended, and so many phænomena by it explained, as of late by that indefatigable and ingenious physiologist, Mr. HUNTER, which has given rise to great controversy. Dr. HENDY has laboured, in his treatise on glandular secretion, &c. to refute all the proofs of Mr. HUNTER, and to destroy the life of the blood, and assigns life and action to the solids only; which doctrine will fully encourage the abuse of the lancet in this metropolis,

in

its power of life; without breathing, circulation is stopped: accordingly we find in *Gen.* ii. 7. first נשמת חיים *neshmat-chiam*, *sufflavitque in nares ipsius halitum vitæ*, and then נפש חיה *nephesh chia*, *sic factus est homo anima vivens*, so that the נפש *nephesh* is the consequence of נשם *neshem*; whereas נפש *nephesh*, denoting growth and motion, are effects of life, consisting in the blood itself, and therefore נפש *nephesh* is always joined with דם *dam* blood, as much as to say, " that the growth and motion of the בשר *basar* fibres " consist in the דם *dam* blood." In another place, I have endeavoured to shew that all the Hebrew words are not accidental or arbitrary, but convey with them the very nature and idea of the thing they denote.

in the hands of the ignorant: an attempt
therefore, to fhew that the blood is alive,
and that we loofe, as it were, by the lofs
of each ounce of blood, an ounce of life,
muft be of the greateft utility to the pub-
lick, and the following fheets cannot be
deemed a mere fpeculation and ufelefs
theory.

Life then has been confidered by fome,
as fubfiding in the fibres, and by others,
as exifting in the blood only; in other
words, life has been annexed either to the
fluids or to the folids of the animal: in
order therefore to decide the controverfy,
it will be requifite (that we may not have
a difpute of words only) to define the
words that conftitute the controverfy,
namely. what is meant by folids, fluids, and
life. But fince we perceive the qualities of
things by their actions only, thefe actions
cannot be confidered by themfelves, indepen-
dent of thofe bodies they act upon, and their
reaction; it will therefore be alfo neceffary

to

to explain the laws of action and reaction, in living bodies, efpecially as it ferves to throw great light upon many phænomena, and as it has not been fufficiently confidered, notwithftanding its great influence upon the practice of phyfic.

Were we to act as empiricks, then certainly the multiplicity of experiments made on living animals, by torturing them, would not only be an infringement upon pity and compaffion, but throw the greateft contempt and fcandal upon phyfiology; in fo much, that the beft phyfiologift, who indefatigably fpends his time in making experiments, would be deemed the moft cruel and mercilefs tyrant, inftead of being defervedly crowned with honor and reputation, for his obfervations that tend to difcover the knowledge of difeafes and their cure.

It would be indeed a mere play of words, and a ufelefs tautology, were the confequence

quence and refult, of all the experiments
made on the blood, only to fhew its dig-
nity, and to honor it with the name of
life; and the controverfy of phyfiologifts
on that fubject, would deferve no more
attention, than the quarrel of children
about a name: but that is far from being
the cafe; on the contrary, the obfervations
on the blood have fo great an influence
upon the practice of phlebotomy, that I
wonder, the ftate of the blood itfelf in
V. S. fhould not be more regarded. The
late ingenious Mr. HEWSON, in his experimental inquiry on the blood, has thrown
out many ufeful and practical obfervations
on that fubject. I fhall therefore only take
notice of what he has either left untouched
entirely, as concerning the red-globules,
or what, as he himfelf fays, he was not
able to explain; and the whole will be
divided into fix fections; 1. of folids,
fluids, and vapor; 2. of action and reaction; 3. of life; 4. of the life of the

blood; 5. of phlebotomy; 6. of the red globules.

I am far from imagining, that the hints which I here throw out will escape all objection, yet I hope they will not be treated with the severity of criticism, but with the generosity of candor, even should some of them be found fallacious; for many true discoveries have been investigated by the means of some new, even false opinions started; and many precious and noble edifices have been raised upon the ruins of others: if that should be my case; if this essay should excite men of real knowledge, and who have more opportunity of pursuing the subjects, by experiments, with more accuracy than my capacity is able to reach, and then either approve or destroy my conjectures; I shall, in both cases, think my labour well paid, and amply rewarded.

E R R A T A.

Page 7. line 16. for *of felway,* read *of the,* &c.

    39. Note *(u)* l. 11. for *tubes,* read *tube.*

    50 l. 2. for *motion is,* read *motion or is.*

    88, Note *(ii)* l. 1. for εχύση υπό, read ἐχύσῃ ὑπό,

    l. 2. for ληφθῆνα, read ληφθῆναι.

    92. Note *(ll)* l. 3. for *which run between the bones,* read *between which the bones run.*

CONTENTS.

SECT. I.

OF SOLIDS, FLUIDS and VAPOR.

Page

ALL bodies appear under either a
solid, a fluid, or vapor   —   1

An objection againſt Dr. FOR-
DYCE's definition of ſolids, fluids, and
vapor removed   —   —   2

The univerſe is kept in motion and exiſt-
ence, by the two oppoſite attractions,
that of coheſion and gravitation. In
the circulation of the blood, two
oppoſite powers are obſervable, the
ſame as in nature in general   3 and 4

Solids, fluids, and vapor, are in a continual
change   —   —   4, 5

b   How

CONTENTS.

Page

How this change may be produced, note *(c)*

Page 5

The opinion of ARISTOTLE is the same
with that of the moderns, only differently expreffed — . — 5

The fituation of the folids, fluids, and vapor,
muft remain undetermined — 6

The folids and fluids in the animal undergo
the fame changes as thofe in nature 7

The great connection between mind and
body — — note *(d)*

The different fecretions in the animal might
be accounted for, if we underftood
the laws of motion and changes in
nature — — 8

Growth and granulation is produced by the
changing of fome parts of the blood
into folid veffels — — 9

Some parts of the animal are continually
evaporated, whilft others acquire a
greater confiftence — — 10

Other fyftems of nature take in what the
animal throws off — ibid.

The operations of nature are fimilar to thofe
of the animal body — — 11

The

CONTENTS.

Page

The fituation of folids and fluids in the animal is undetermined —— 11

Mixture the delight of nature —— 12

The effect of mixtures in medicines —— 13

Theory is clofely combined with practice 14

SECT. II.

OF ACTION and REACTION.

To perception and motion, both action and reaction are neceffary —— 15

Why the fame objects will operate differently in different circumftances 16

The great concern the law of action and reaction has with phyfic —— 17

The phænomena of fevers explained by action and reaction —— 18

The operations of medicines by the laws of reaction — — 19, 20

No inference to be drawn from the operations of medicines, in a found ftate of the body, to their effect in its difeafed ftate — — — 21

All pleafure and pain excited by reaction 21, 22

Why fometimes weak, and at others ftrong

parts,

CONTENTS.

Page

parts, will be more affected by applications — — 22, 23

SECT. III.

OF LIFE.

What bodies are called dead, and what
living — — — 25, 26
The ftriking properties of living bodies, 27, 28,
29
There is either a repulfive or affimulating
power in the motion of bodies — 30
Life claffed under three general kinds — 31
Three kinds of motion — — 32, 33, 34
Uterogeftation a procefs of life — 35
The fear of the inferior animals to the fupe-
rior, by the power of life — 35, 36
The different gradation of life — note (t)
Living animals produce both cold and heat
Page 37, 38
Different fubftances converted into one and
the fame nature, by the power of
life — — 38, 39
In what manner medicines may operate
note (u)
Digeftion a power *fui generis* — Page 41
Living

CONTENTS.

Page

Living fyftems preferve themfelves without
exertion —————— ———— 42

SECT. IV.

OF THE LIFE OF THE BLOOD.

Life exifting in fluidity and motion —— 43
Why the blood may be called alive —— 44
The phænomena of the blood's remaining in
    a fluid ftate when in, and coagulating
    when out of the veffels, are accounted
    for by its living principles, 45, 46, 47, 48
Dr. HENDY's objections to the proofs of
    Mr. HUNTER examined ———— 48
The vigor of vegetation confifts in the juices
    of the tree and in its life — 49 to 60
The fecond proof of Mr. HUNTER —— 60
The third proof of Mr. HUNTER and DU
    HAMEL's cafe explained —— 62, 63
The adhefive inflammation occafioned by
    the furrounding living fluids —— 64
An obfervation of Dr. HUNTER ——— 65
The principle of coagulation owing to the
    irritability of the blood —— 67
The fourth argument of Mr. HUNTER — 68
The fifth argument, and the difference be-
                tween

CONTENTS.

Page

tween the coagulation of a jelly and
    that of the blood, and the fixth ar-
    gument   —   ——   —   69, 70

Any living animal dies the fooner, the
    ftronger it reacts   ————   71

A difference between irritability and weak-
    nefs   ————   ————   72

Whether the blood coagulates only by
    ftimuli   ———   —   73

The feventh argument   —   —   74

A difference between the anaftomofation of
    nerves and that of blood veffels,   75, 76

Each fet of nerves has a particular office, 76, 77

The laft objection of Dr. HENDY removed   78

S E C T. V.

OF PHLEBOTOMY.

A difference between a real plethora and
    a fudden repletion, occafioned by
    repeated V. S.   —   —   79

A general relaxation will be produced by
    repeated V. S.   —   —   80

How the ftate of the fibres may alter the
    properties of the blood,   81, 82, 83, 84

The fizy cruft a mifleading indication of
    V. S.   —   —   —   —   85

The fizy cruft is the fymptom, not a difeafe
    itfelf   —   —   —   78

The

C O N T E N T S.

Page

The danger of repeating V. S. in pregnant
   women   ——   ——   88
It is better to bleed a little, and repeat it,
   and pay attention to the ftate of the
   blood, than draw off a great quantity
   of blood at once   —— . ——   89

S E C T. VI.

OF THE RED GLOBULES.

The red globules a compofition of earthy
   and oily fubftances   — . 90
The two different offifications between
   membranes and cartilages explained, 92,
                                                   93
A fuppofition that offification goes on continually, even after growth   —— 94
Gravelly and calcarious complaints, occafioned by a fault in the offification,
   owing to the ftate of the red globules   —— . —— 95, 96, 97
The pain in the gout explained   — — 99
Some difeafes may be produced by the
   want, and others by the abundance
   of the red globules   — — . 100

Published by the fame AUTHOR,

A

D I S S E R T A T I O N
ON THE
L A W and S C I E N C E S;
In HEBREW.
Q U A R T O.
Price FIVE SHILLINGS.

———————————

T H E

The SPIRIT and UNION of the
Natural, Moral and Divine Laws.

Price Two Shillings and Six-pence.

O N T H E

B L O O D, &c.

S E C T. I.

OF SOLIDS, FLUIDS, AND VAPOUR.

SOLIDS, fluids, and vapour, are the three moſt ſtriking objeꞓts we perceive in all bodies, and in nature; every ſubſtance appears under one or other of theſe three. The moſt correꞓt and general definition of ſolids, fluids, and

B   vapour,

vapour, is that of the accurate and learned Dr. GEORGE FORDYCE; *(b)* namely, bodies, whofe attraction of cohefion is ftronger than that of gravitation, are called folids; fuch bodies whofe attraction of gravitation overcomes that of cohefion, are called fluids; bodies which can be condenfed into fluids, are called vapours.

There has been an objection raifed againft this definition; namely, that it is giving only the caufe and the means by which bodies are folids, or fluids, not the real diftinct figns of the appearances; which fhould always be the aim and ufe of every definition; but it has perhaps not been fufficiently confidered, that by attraction is meant an effect, or the appearances of bodies with refpect to their relative fituation; which appearances are objects of fenfe, not the power or caufe by which that attraction or effect is performed;

(b) Chymical Lectures.

formed; the cause itself being occult, may be an impelling one: now then, according to the attraction of cohesion, the solidity of a body will be encreased; as the contrary effect, namely, fluidity, will be produced by that appearance, called attraction of gravitation prevailing over that of cohesion; which is the tendency of each particle of a body, to another center, and as it were tearing itself from the center of that body to which it belongs. By these two opposite attractions, or constant action and re-action, the universe is kept in motion and exiftence. How these opposite motions are produced alternately, I do not pretend to determine; sufficient it is, that such motions are perceptible in the universe, and its parts.

In the circulation of the blood, we observe, that the power which produces the motion of the lacteals, veins, and their contents, in bringing the chyle and blood to the right auricle and ventricle of the heart, cannot be equal to that motion

with

with which the blood is expelled again from the left ventricle into the *aorta*, to all the extremities; nor can either of thefe two motions be fo ftrong as entirely to overcome the other; for in both cafes, circulation would ceafe; each motion therefore muft always, by fome means or other, probably by refpiration, alternately gain a new force to keep up the circulation: the fame may hold good of the motion in the planetary fyftem and their particles, as in each particle of the blood and veffels.

In confidering folids, fluids, and vapour, we fhall often find them fo changeable and blended, as not to be able to determine either their fituation or nature. The conftant alternate motions of the planets, or the two different attractions, produce the continual change of fluids into folids, folids into fluids, and thefe into vapour,

and

and the reverfe; *(c)* thus fome bodies, which, while they remain in the bowels of the earth, are folids, when exhaled in the form of vapour, become fluids in the atmofphere, which is very perceptible in moft of the acids, &c. ARISTOTLE, though not well acquainted with the real properties of bodies, yet feems to teach this doctrine, when he fpeaks of his four elements, earth, water, air, and fire, as changing continually into one another; moderns maintain the fame, though they exprefs it in other words, when they fay, that " the different pro-

" perties

(c) Which may eafily be underftood, if we fuppofe any aggregate of infinite fmall elaftic particles, preffed continually upon a certain center, and repelled again from it, that by that motion fome of the fmall component parts, are always preffed out of that aggregate in the form of very fmall particles, fo as to affume the appearance of fluids and vapour; whilft others, which are fqueezed towards the center, acquire a greater confiftency and folidity, by the force of cohefion, and take on the appearance of folids.

" perties of bodies are owing to the
" different combinations of their integrent
" parts."

If we confider the fituation of folids and
fluids in the univerfe, we hardly can determine, whether the fluids run through
the folids, or whether the folids are
plunged into the fluids: thus, according
to one appearance, we may fay all the
folid planets move in the air or æther; and
with refpect to another appearance, we
may, with the fame propriety, fay that
the æther or air is diffufed through them :
but if we take both appearances together,
we fhall find the folids and fluids fo united
and blended into one another, that we
fhall be at a lofs, which of the conditions
to afcertain; and we fhall find that there
àre no real folids nor real fluids to be found;
and that we only ufe thefe terms to diftinguifh bodies according to their different
appearances.

If

If we apply this reasoning to the animal body, we shall find that its solids and fluids are equally united, and undergo the same changes as those in nature at large, so that no alteration in the animal body can be accounted for by the state of its solids, or fluids only; *(d)* and I do not doubt, but that

------

(d) Not only the apparent solids and fluids in the animal œconomy are closely connected and reciprocally affected by each alteration; but also the mind is closely connected with the body, so that many diseases of the stomach have their origin from the situation of the mind, and the mind again is affected by the circumstances of the stomach. A head-ach will very often be cured by an emetic; a chearful mind will increase the appetite, and the digestive power. In consequence of certain ideas being excited in the mind, an increased action of the nervous influence, will follow upon the muscular fibres of the stomach, and it will be affected with nausea and vomiting; and that idea will stimulate the stomach the same as a dose of *ipecacuanha* or emetic tartar; the sight or remembrance of grateful food, will increase the secretion of salivary glands; all the hypochondriac and

hysteric

that if we underſtood rightly the laws of
motion and changes of nature in general,
we might eaſily account for all the different ſecretions and motions in the animal
body : for which reaſon the ancients
called man a microcoſm. Each part of
the animal body is in a perpetual, though
inſenſible motion, by which motion it
converts

---

hyſteric ſymptoms are often the effeꞔts of an uneaſy mind,
and when they are the effeꞔt of a ſedentary life, the mind
is always ſecondarily diſturbed. Thus then the ſtomach,
which is the firſt means of affording nouriſhment to the
body, and generating all the groſſer parts of the fluids,
and the mind, which is the firſt elaboratory of the finer
parts of ſolids and fluids, are ſo blended into one
another, that we need not wonder when diſeaſes are
miſtaken, by ſearching for their cauſe in the one,
whilſt it is ſituated in the other : and we may add, that
every phyſician ſhould alſo be a metaphyſician, that is,
he ſhould underſtand the mind and its diſeaſes, as well
as thoſe of the body. In another place I have endeavoured to ſhew how both ſciences, phyſics and meta-
phyſics, are one and the ſame : *Vide*, the Spirit and
Union of the natural, moral, and divine Law. Page vi.

converts continually some solid parts into fluids, and these into vapours, and the reverse; the former is evident from all the different evacuations and insensible perspiration, and the latter from growth and granulation: that growth is effected by the change of the parts of the blood into the solid vessels, is evident from the great quantity of blood which is found in the arteries, as long as growth continues; whereas in old age more blood is found in the veins; granulating flesh can be nothing but a mass of fluids changed into solid parts, and as Mr. HUNTER justly supposes, the coagulable lymph. The coagulable lymph itself, which may justly be called a solid, took its origin from a fluid, the chyle. Air taken into the lungs is mixed there with the blood, alters its nature, and becomes more fixed than respirable air: the food, which is carried down to the stomach by the contractile muscular power of the œsophagus, is by a similar exertion of muscular power, carried

from

from the ftomach againft its gravity,
through the different convolutions of the
inteftinal tube, and conveyed out of the
body; while, by a fimilar motion of
the fmaller veffels in the mefentry and inteftines, the chyle is carried to the thoracic
duct, from thence to the veins, where it
acquires a greater confiftency, and from the
veins into the arteries. By that conftant
motion and change, fome parts of the animal body are continually evaporated internally and externally, whilft others acquire
a greater confiftence; which makes it very
probable, that fome parts of the blood are
changed into parts of veffels, and parts of
veffels into the form of vapour.

All the parts, which are thrown off
from the animal body, are taken in again
by other fyftems of nature, (with the fame
motion as they were at firft taken into
the animal body) and become parts of
them; where they undergo various changes,
'till they are thrown off again from thefe
fyftems,

fyftems, and affumed by others, and fo on : by thefe means nature in general carries on life, while its parts or individuals are firft generated by the whole, and then, as it were, by it devoured again; for which reafon, ARISTOTLE in defining the world faid, " It is the con- " ftant production and deftruction of " things."

Thus then the operations of nature through the whole, are fimilar to thofe of the animal body upon its parts.

Again, with refpect to the fituation of the folids and fluids in the animal body, it muft remain undetermined, the fame as in nature; fince I may fay, according to one appearance, the folids are immerfed into the fluids of the animal; becaufe the veffels, and each of the *vafa vaforum*, are not only filled with fluids, but the whole body is furrounded with its own vapour,

and

and has its own fluid atmofphere; and, if I confider another appearance, I may reverfe the pofition; but taken both together, we fhall find, as in nature, that no real determination of the permanent fituation of folids, fluids, and vapour, can be fixed; and that thefe three principles run continually into one another, and each of them is in conftant motion, and continually changing.

Thefe three principles, folids, fluids, and vapour, are in nature mixed together; it is mixture in which, as it were, nature finds delight, by which it produces and deftroys continually. There is nothing perhaps fimple in nature, nor do thefe objects which appear fimple, affect either the animal body or the mind, in fuch a degree as compounds do. We are foon tired of the fimilarity of objects; and, concerning the paffions, I have fhewn in another treatife,

treatife, *(e)* that mixed paffions are the ftrongeft. The learned Dr. FORDYCE obferves, *(f)* that mixtures of medicines are generally both more falutary and efficacious in their operations. *(g)* Dr. LEWIS, in his materia medica, relates that he found fifteen grains of jalap, mixed with two grains of ipecacuanha, proved more efficacious than twice that quantity of jalap by itfelf; and I have found the fame in the mixture of alteratives, deobftruents, or relaxants (as fome moderns call them) the trial of which is recommended by Dr. FORDYCE in his lectures; and indeed many

other

(e) Vide, the Spirit and Union of the natural, moral, and divine Law. Page 203.

(f) In his Lectures on the Materia Medica.

(g) In the conducting of this part of medicine, Phyficians have always run to one of the extremes; the ancients mixed millions of medicines, whereas fome moderns prefer mere fimples.

other mixtures might be tried with fuccefs:
all which proves alfo, how clofely theory
is combined with practice: we find in
theory, and in confidering nature in general, what powers are produced by mixture,
and we find the fame in the practice of
phyfic.

SECT.

SECT II.

OF ACTION AND REACTION.

ALL animals are set in motion or excited to perceive, and men to think by the reciprocal action and reaction between them and surrounding objects; because in order to perceive, or have an idea, it is not only necessary, that external objects should operate upon the animal, but that the animal also should operate in some measure upon them, otherwise perception is hindered. Thus if some objects are represented to our eye, while we are engaged with others, we do not perceive them at all; it is the same with sounds upon the ear, tastes to the tongue, &c. Whatever excites the animal to any action is done by cer-

tain

tain properties of the objects ; which properties act by certain laws according to
the properties of the animal they operate upon, and therefore we shall often
find, that one and the same application
will operate differently upon different a-
nimals, nay even upon different individuals
belonging to one species, it will even differ in the same animal under different
circumstances, all which must be accounted for by the dispositions of the
animal, and its reaction. Thus will some
harmonies excite different passions in different animals, or in the same animal at
different times : a colour, which seems
pleasing to one animal, will be disagree-
able to another, the same with respect
to taste, smell, &c. We find this di-
stinction even in the effect of mere ideas ;
a fright, for instance, will in one person
encrease the secretion of several glands,
stimulate the intestines so as to discharge
their contents, and in others it will have
just the opposite effect ; all which is owing

ing to that great general law of nature, I mean action and reaction. In order then to perceive any subject rightly, both action and reaction must be proportioned. If the reaction be stronger, the object will not appear at all, as is the case in small objects, but when proportioned they will as it were unite, and the intention of nature will be performed. This law has so great an affinity and concern with physic, that not only all the physiological changes in the animal body, but even all its alterations in a diseased state, the application of medicines, its experimental philosophy, &c. cannot be adjusted without considering the reaction. *(b)* Thus

C   the

(b) Dr. Hendy, in his treatise on glandular secretion, concludes, " that pus does not act as a ferment; " because a piece of mutton which he introduced into " an ulcer of a leg producing pus, was not converted " into pus"; forgetting that pus acts as a ferment only in living parts by their reaction, and that a slice of mutton may be considered as a piece of wood, both being dead parts, not affording any reaction.

the proximate caufe, or the firft ftriking appearance of almoft all fevers, is a debility, which is a ceffation of motion or languor, and which is, or by which the cold ftage is produced, agreeable to what we perceive in nature in general, where cold and reft are coexifting qualities; the denfer a body is, that is to fay, the lefs inteftine motion its integrent parts poffefs, the colder it is, which is evident in different minerals and ftones. Upon the cold ftage of a fever, the hot paroxyfm follows by the reaction and encreafed motion of the veffels, and their contents, which paroxyfm, or reaction, is always proportioned to the action or the caufe of the fever, or to the cold ftage, to the degree of tremor and the degree of the debility; the anorexy naufea vomiting, in the cold ftage, is alfo owing to the reaction of the ftomach and the internal veffels, refifting the preffure made upon them by the contraction of the external veffels. In this manner we may ratio-

nally

nally account for many other fymptoms. *(i)*
The fame law of re-action holds good in
the operation of medicine. *(k)*   The fame
medicine ˋ

(i) An emetic exhibited in the beginning of a fever,
even putrid, *(vide,* Sir J. Pringle's Obfervat.
pag. 306.) will finally put a ftop to it, by exciting
the re-action of the fyftem, either to expel the morbid
matter, or to remove the contraction; in the fame
manner will antimonials, fo given as only to excite a
continual naufea, prevent the paroxyfm of fevers, by
flightly re-acting the morbid matter, or the contraction
which infinuates itfelf gradually into the fyftem, and at
laft produces the paroxyfm; and I fhould imagine, that
medicine in general act only by exciting fome re-action
of the fyftem, to expel, remove, or counteract the
difeafe; not that there is any fpecific virtue in any
medicine itfelf.

(k) Opium will, in fome circumftances of difeafes
act as a corroborant, procure fleep, and haften the cure;
(vide, the learned Sir J. Pringle's Obfervations,
page 132) whereas it will have quite the contrary effect
in other difeafes, and in a found ftate; all which muft
be accounted for by the re-action of the fyftem, not by
the action of the opium alone. And hence the hurtful
confequences of all unneceffary medicines, by exciting

a preter-

medicine will produce fometimes quite
the oppofite effect in a found ftate of body,
to

---

a preternatural reaction in the fyftem, and making
it, not only, as it were, at the fame time neglect its
natural actions, requifite to preferve health and life,
but produces alfo, by that preternatural exertion, a
future debility, which renders the fyftem unfit to continue its natural functions. Some phyficians are hence
led to conclude, that nature muft cure all the difeafes,
and that medicines are entirely ufelefs; but they forget
that our prefent manner of living is not a natural one,
and that we are no more the natural man; fo that we
muft often have our recourfe to art, in order to re-act
the difeafes produced by art: animals and favages who
live entirely according to the dictates of nature, ftand
feldom in need of medicines, phyfical evils being very
few; but men, civilized by art, have broken, as it were,
their alliance with nature, by oppofing it continually;
and therefore, when a difeafe takes place, they muft
firft expel the enemy by artificial application of medicines, before they can expect the cure of nature.
Vitiated demands may often be miftaken for calls of
nature; under certain circumftances, there will fometimes be a falfe appetite, owing to a difeafed ftate of
the ftomach and inteftines; if that call be fatisfied, the
difeafes will only increafe; but in this and the like
cafes,

to what it does in a difeafed ftate, and fo in different difeafes; hence the miftake of fome, who, by trying the effects of fome medicine upon themfelves in a found ftate, would infer the fame to be their effects when applied to a difeafed body.

It is, I believe, beyond doubt, that we acquire our firft ideas by refiftance; we are refifted upon our fenfes by all external objects, in our firft entrance into the world, and we endeavour to perceive the things that refift us, and to react upon them; were we merely paffive, without reacting, we could have no ideas. All pleafure and pain are excited, and felt more or lefs, according to the degree of reaction, which degree of reaction con-ftitutes the degrees of pain or pleafure.

If

cafes, art is wanting to remove the caufe of that falfe appetite, and then certainly nature will finifh the cure; but not that nature is to be left to itfelf, when it has been hindred in its functions by corruption or art.

If harmonious founds ftrike my ear, when my attention is taken off, no pleafure will enfue, nor will pain enfue under the fame circumftances, by difproportioned and noify founds. The pain which we feel from cold applied to any part of the body, is excited by the preternatural exertion of the part, to produce heat and react the cold; the pain will increafe the more we think on it, that is, the more we endeavour to react it.

The cutting of a ftrong mufcle will be far more painful than that of a weak one, the former being ftronger, and confequently reacting more to the cut, than the latter, and therefore, when the mufcle is once cut through, and confequently no more reaction, the pain ceafes entirely; there is now and then to be obferved, that fome applications on weak and irritable parts will caufe great pain, whereas on ftrong parts, no fuch fenfation will enfue; but even that muft be accounted

for

for by the reaction. A flight touch on a
ftrong part will not be fenfible or refifted
at all, and therefore will not excite reaction; but that very touch to a weak or
irritable part will threaten its deftruction,
and will be fo fenfible to it, as to excite
reaction; irritability itfelf is nothing elfe
but the reaction of fenfible parts to any
refiftance: moft membranes are irritable.
The influence of the imagination in thefe
cafes, which we know may diminifh or
increafe the pain, is alfo to be accounted for
by the reaction. A blifter applied to a
found part, will caufe inflammation and
great pain, which is certainly owing to
the reaction of the parts; fince the fame,
when applied to a relaxed or paralytic
part, will not have the fame effect. Pain
itfelf, which attends all inflammation, is
owing to the reaction of the fmaller veffels
to the blood, which is continually impelled on them; and therefore, by relaxing
the parts inflamed, and taking off that

tention

tention and reaction, an inflammation will often be cured.

All application of medicine act according to the reaction of the parts they are applied to; and there are no medicines, strictly speaking, possessing any quality not relative to the parts and their circumstances, they are applied to.

S E C T. III.

O F   L I F E.

ALL bodies are diftinguifhed, and confequently differently named by the different impreffions they make on our mind, according to their properties; we call thofe bodies alive, which ftrike us with properties peculiar to themfelves, and not to be found in other bodies which we call dead.

Bodies, therefore, that poffefs fome properties, by which they are diftinguifhed from dead bodies, we call living. We fay vegetables and animals are alive, becaufe they have qualities which do not appear in minerals: to denominate any thing alive, we muft difcover thofe pro-

perties,

perties, in which it differs from minerals, and correfponds with vegetables or animals; the more properties of the latter it poffeffes, the more it has right to be ranged under the clafs of living bodies: our firft bufinefs is therefore to point out, and enumerate all the properties, by which bodies are faid to poffefs life, and then to fearch for thofe properties in bodies, the life of which does not ftrike us immediately, and if we find them poffefs fome of the living qualities, we need not hefitate to call them in fome fhape alive.

The moft ftriking properties of bodies we call alive, are the following: 1ft, *Self Motion. (l)* All dead bodies continue at reft 'till they are impelled by others, when they begin

---

(*l*) Inanimum eft omne quod pulfu agitatur externo; quod autem eft animal, id motu fcietur interiore et fuo. Nam hæc eft propria natura animi atque vis: Cicero Difput. Tufcul. lib. 1.

begin to move in fuch a direction, as they are impelled to; but bodies alive, poffefs a continual inteftine motion within themfelves, without the impelling caufe being fenfible: I fay fenfible; for I think there muft be a continual impelling motion, to produce the motion of living bodies, though the caufe may be infenfible. 2d, The *confent* and *co-operation* of the parts, to promote the continuance and health of the whole; as foon as that co-operation is loft in any part of the fyftem, a difeafe is produced; it is by that pro-perty, we call a mufcle alive as long as it contracts alternately, by the applica-tion of a ftimulus, even after it is fepa-rated from the living body; becaufe it con-tracts by the confent of the whole, though only one part of it is ftimulated. *(m)* By

the

(m) The caufe of that kind of life, is a matter of difpute. HALLER thinks it is owing to the glutinous

matter

the fecond quality, a third is produced, that of *affimulation*; living bodies continually take in other fubftances, and render them homogenous to their fyftem. A fourth is alfo produced; that of *refifting* and *reacting* to any heterogenous body that threatens to deftroy it. A fifth quality, owing its origin alfo to that of confent and affimulation, is that of the *conftant converfion of folids into fluids, and fluids into vapour,* and the reverfe: thus folid food taken into the body, is converted by the power of digeftion into a fluid, the chyle, and blood, &c. After fome parts of thefe fluids have nourifhed the body, others are exhaled, both internally and externally,

in

---

matter connecting the earthy particles, of which the fibres are compofed, which opinion has been refuted by Dr. WHYTT, in his phyfiological effays; but I fhould imagine, we never fhall be able to determine caufes, which are not objects of fenfe, and we fhould only take notice of the effect, which is juftly called by modern phyfiologifts, a living principle, to diftinguifh it from the effects of dead bodies.

in the form of vapours: fluids which are taken in, are changed into a more folid form, the chyle; that again is converted into blood, a part of which, the coagulable lymph, is a folid; and it is very probable that the folid parts of the blood veffels themfelves are infenfibly formed from the blood, namely, the coagulable lymph, whilft others are thrown off. The change from fluid parts into folids is evident enough, from the circumftances of growth and granulation, and from every appearance of nature, fimilar to that change in the univerfe where we fee that continually fome parts of the ocean are changed into parts of the earth, whilft many parts of the earth are converted into the ocean.

All the above mentioned properties and powers are done by motion; the motion of bodies therefore feems to be of fuch a nature, as either to repel other bodies which come in contact with them, and

are

are heterogenous, or to unite with and
affimulate them, that is, to render them
homogenous. The repulfive power is e-
vident in all natural bodies in general;
and that power of affimulation, in the
vegetable and animal fyftem in particular.
Vegetables and animals when in contact
with bodies which they cannot affimulate, endeavour to repel and deftroy them
by reaction, (fect. 2.) and when not ftrong
enough, they are, by their own reaction,
(fect. 2.) and the action of thofe bodies,
deftroyed. Thus is food taken into the
ftomach and affimulated by the whole
fyftem, but if any thing enters the ftomach which tends to deftroy it, the ftomach endeavours to react and throw it
either out, or is, by that very reaction,
and the action of the heterogenous body,
deftroyed. All this is done by motion
and life; by motion, I mean the external appearance of the body with refpect
to its place; and by life, the internal
power which promotes that motion.

Life

Life may be claffed under three general kinds; 1ft, the life of a fyftem in
general, as that of an animal or vegetable; 2d, the life of its parts when taken
from the whole, as a mufcle, a twig;
and 3d, volition, or the foul, which laft
differs according to the different organization of the animal, *(n)* and is rather
the fubject of metaphyfics, which I have

en-

(n) I would not be underftood here to hint at the
repugnant doctrine of *materialifm*; on the contrary,
I cannot imagine how the certain fituation or ar-
rangement of the various parts of the body fhould be
able to produce fenfe or reafon. " *Membrorum vero*
" *fitus et figura corporis, vacans animo, quam poffit*
" *harmoniam efficere non video.*" *Cicer. Tufcul. Difput.*
Lib. 1. We certainly muft fuppofe the exiftence of a
mind, but neverthelefs that mind ftands in certain unknown relations and connections with the organization
of the body it enlivens and perceives ideas according
to the nature of the mechanifm in the organs of the
fenfes, by the impreffion and action of the furrounding
bodies upon them, and their reaction (fect. 2.), which
has given rife to LIBNEZE's doctrine of the pre-efta-
blifhed harmony.

endeavoured to explain in another place. Accordingly there are three kinds of motion; 1ft, the motion of the fyftem, or original motion, which is nearly circular ; (o) (fect. 1.) all its parts repelling equally from

---

(o) By circular motion I mean, that there is a certain center in the fyftem, which may be called the grave center, to and from which all the parts move continually, and confequently form by that motion a circle : that motion when it is, by fome caufe, inverted, fo that thofe parts which fhould move towards the center and be affimulated, move from it, and others which fhould move from the center and be exhaled, are impelled towards it, then convulfions take place, fuch as are produced by a fright, or by the application of ftimuli, which, by exerting the parts to react, confufe the natural motion from and to the center. That fome ideas are capable of fetting the whole fyftem in a convulfive motion, is not furprifing at all, fince we fee the fame effect of the will in a found ftate; I mean the act of jumping, which is done by the will fetting the grave center of the fyftem, the pelvis and its mufcles in motion to a certain direction, fo that the whole body is fet in motion and is lifted up.

The

from all fides in an alternate motion, to and
from the center; fuch are the motions of the
planets, animals and vegetables; which
motions are original, *(p)* and always con-

D

nected

The vermicular motion of the inteftines, to prefs
downwards the excrements, which is made by the contraction of the fpiral fibres, and hence called peri-
ftaltic, is not peculiar to them; for the motion of the
blood veffels feems to be of the fame nature, by contracting continually their fpiral mufcular fibres, and pro-
pelling the blood. The ftructure of the different coats of
the blood veffels, which are fibres, fpread longitudi-
nally and latitudinally, when fet in motion, form evi-
dently that kind of vermicular motions which is fo fenfible and evident in the inteftinal tube; and I fhould
be apt to conclude, that all motions in the animal
fibres are of the fame nature; fome particles move to,
and others from the center in all directions, and form
both a circular and vermicular motion.

(p) When I fay original motion, I mean a motion
where its caufe is infenfible, or is not an object of fenfe;
for metaphyfically confidered, there muft be fome
caufe even for the original motion which depends on
life, and that caufe muft communicate the motion, or
repel it conftantly.

nected with the continual change of the
three principles, folids, fluids and vapour:
(fect 1.) 2d, the motion of parts; and
3d, the invifible motion, or that of the
mind. *(q)*

Many other phænomena appear in living
fyftems not eafily accounted for by mechanics, fuch as for inftance, that power by
which the fyftem is firft produced, and
then generated by feeds. The famous Dr.
HUNTER, in his anatomical and phyfiological lectures, fhews by many proofs, that
utero-geftation is not done by any mechanical power, and that the uterus is not dilated
by

(q) That the mind performs its operations, thinking,
concluding, &c. by motion, no one can doubt; becaufe
thought is the conftant prefentation of ideas which are
acquired by objects; by the images of which we think,
and which images we conftantly move in our imagination, as it were, from one place to another, by compa-
ring fome and finding differences between others. (See
the fpirit and union of natural and moral divine law.
Lect. 10.)

by the bulk of the fœtus; one of his proofs
is very ſtriking, he found the uterus dilated
when the *corpus luteum* and the impregnation was only ſeen in the *ovaria* : he affirms
therefore, that it is a proceſs of life *(r)*
which dilates the uterus, and produces the
placenta and its veſſels for the nouriſh-
ment of the fœtus.

It is by that power of life, that the
inferior ranks of animals are afraid of
the ſuperior, which cannot be owing to
figure or bulk, ſince as ſoon as a man is
dead, that fear is gone *(s)* though his
figure

(r) This proceſs may be owing to the uterus being
ſtimulated by the impregnation; the ſtimulation cauſes
a determination of blood to the uterine veſſels, as a
proviſion and habitation for the fœtus, and enlarges the
ſize of the uterus during pregnancy, its veſſels being
diſtended and filled with blood.

(s) Rabi Simon, ſon of Eliesar, ſays " A living
" child, of one day old, guards itſelf from the mice;
" whereas

figure remains the fame; and many animals, of greater bulk than man, are afraid of man, which fear feems to be excited in them by the higher degree of life man poffeffes. *(t)* It is with that power

---

" whereas Og the King of Bafan (the greateft giant) " muft be guarded from them;" becaufe it is faid, *Gen.* ix. 2. " And your fear, &c." as long as a man is alive, his fear lies upon all creatures; but as foon as he is dead, that fear vanifhes. Talmud. Shabatt. Sect. Hafoel. fol. 151.

(t) I muft remark here, that though the gradation, obferved in the univerfe, from the loweft clafs of minerals, to the higheft clafs of animals, men, is fo obvious, that it has ftruck all the ancients, the Jewifh writers, the Kabalifts, and PLATO efpecially, who have extended it fo far, as to trace by analogy, other beings, as angels, &c. to the omniprefence himfelf; yet, I think there is a material and effential difference between different beings, though they may be gradually traced as a chain. That very thing or quality, which diftinguifhes one being from another, makes it alfo effentially different: vegetables are faid to be alive, yet there is the moft material and effential difference between

the

power that the animal produces, as some moderns express themselves, both heat and

the life of the higher class of vegetables, and that of the lower class of animals; and so there is between the higher class of animals, and the lower class of men, as there is also between each degree: we discover a similitude between different beings, and a chain in the whole creation; we find the same quality, which the lower degree of beings possess, is also connected with beings of a higher degree; but that very quality, which we think they have in common, must be essentially different in both, its being in the higher degree combined with another, with which it could not be, were it not quite different; and why should not the lower degree possess it also, were there not a material difference between the nature of even those qualities they are said to possess in common? Thus vegetables grow, as well as animals, but there is a material difference between these two kinds of growth; were they the same, both would either possess the same kind of life and sense, or none of them would be sensible: the lower class of animals have the sense of feeling or touch, as well as the higher class, but there must be a material difference between that very sense in both: the higher class of animals possess, in some respect, reason; but I think there is an essential

difference

and cold, according to the furrounding atmofphere it finds itfelf in. In great heat, which
tends to deftroy it, by increafing the heat and
motion of the internal and moft material
parts, it throws then the blood from the internal upon the external parts, and produces
cold, by a diminution of motion; and when
in a cold atmofphere, it throws the blood
internally, and increafes there the motion,
and produces heat, in order to defend thofe
material parts.

It is life that converts different fubftances of different properties into one and the
fame nature, as well when fecreted by the
different glands, as by the ftomach and inteftinal tube. Mr. HUNTER fed fome dogs
upon

---

difference between the reafon of an animal and that of
men; and fo, if there are more fenfible beings gradually from man to God, there is no doubt, that the
laft and the moft compleat creature, is effentially different from the creator himfelf. The Hebrew reader
may fee the firft chapter of Maimonedes' 8 philofophical
chapters.

upon vegetables, and others upon animal
food only; the milk of both was analyzed by Dr. FORDYCE, who found them
the fame in their chymical properties.
(*u*) Mr. HUNTER found also, the fame

of

(*u*) It is agreed by all phyfiologifts, that foffils are
indigeftible, fo are likewife all effential oils: both retain
their natural qualities, without fuffering the leaft
change from the procefs of digeftion; in colours, which
may owe their origin to the foffil kingdom, we obferve the fame, as is evident, from the tinging of bones
in animals fed upon madder, and from the colour of
urine, after the exhibition of fome medicines, as rhu-
barb, &c. The undeniable ftriking effects of the bal-
famics, whofe operation on the lungs is done after they
have paffed the procefs of digeftion in the inteftinal tubes,
and the furprifing effects of the bark, opium, &c. are
fufficient to vindicate the reputation of medicine, which
has fuffered much by that new difcovery of the powers
of digeftion. The experiments made by the ingenious
Mr. HUNTER, however, are fo convincing, that it
requires great confideration to determine in what manner
medicines are able to operate. Some are of opinion,
that the medicines exert their operation, whilft in
the ftomach, before they are digefted, by the fym-

pathy

of the cream in the *duodenum*, though the animals had been fed upon different aliments;

---

pathy the ftomach has with the whole fyftem; they make this opinion probable by many examples, where, as foon as the medicine enters the ftomach, it fhews its effect on fome other parts, or glands of the fyftem; fee Dr. ALSTON's Materia Medica, on the Cort. Peruv. and Dr. WHYTT on Opium. But the operations of aromatics and cathartics upon fucking children, when exhibited to the nurfe, are undeniable proofs, that thefe medicines exert their effects, after they have entered into the blood. Perhaps all the operations of medicines confift in their foffil part; the furprifing effects of iron, antimony, mercury, arfenic, fulphur, lead, and all mineral acids and alcalis, are fufficient to make this opinion probable: vegetable acids themfelves may be mineral acids, modified in the organization of the vegetable, which is more probable than to imagine a new acid originally in the vegetable. I am the more inclined to attribute the power of medicine to the foffil part, fince it is well known that hufks and bark are not digeftible, and fince even iron has been found in the blood; the component parts of the fibres are hitherto unknown, and may not they be of the foffil kingdom? So that fome foffils and

minerals,

ments; he found even the fame pure cream, when the animals were fed upon putrid meat, which is alfo a procefs of life to refift and convert putrefaction, evident in many of the lower clafs of animals, as maggots, &c. which live in and upon putrid fubftances; fo do all vegetables, (fee Dr. PRIESTLY on Air) by which Mr. HUNTER fhews, that digeftion is a power *fui generis*, and not a kind of fermentation, fince it converts putridity into a nourifhing chyle; and on the contrary, whenever the power of digeftion is weak, and the food undergoes, by the heat of the ftomach, its own acid fermentation, then eructations take place. It is by the power of life, that living animals

minerals, when exhibited in difeafes of the folids, may perform their effect by incorporating themfelves into the fibres, repairing or renewing them. Were the mother, I mean the ftudy of foffology, as much courted by phyficians, as her daughter, botany, many ufeful difcoveries might perhaps be made: fee alfo note *(i)*.

mals refift digeftion; worms, when in the inteftinal tube, make their nidus as it were in it, inftead of being digefted or deftroyed by its motion. It is by the power of life, that fyftems preferve them-felves without exertion, as we find fome animals lying dormant for a confiderable time, during the winter, without dying; and by the fame means, all vegetables preferve their life, during the winter, elfe they could not revive again, as it is commonly expreffed, in the fummer, though all thefe properties may be pro-duced mechanically, or rather chymically, in a manner not known to us, by the fubtile ftructure of the animal, and by its different motions and directions, when alive.

S E C T.

S E C T. IV.

OF THE LIFE OF THE BLOOD.

IN the preceding section, we have explained that by the word life, is meant that property of a thing, by which it diftinguifhes itfelf from a dead part; we faid that a mufcle is called alive, while it has the power of contracting and dilating alternately, by any ftimulus which a dead part has not; we may then fay, that life confifts in motion, and death in reft; and the reverfe, namely, the motion of a body is always according to the quantity of its life, and that again is always proportioned to the fluidity of the body; parts the moft fluid are poffeffed of the greateft motion, and the greateft power of life: we find accordingly, that the fluids

are

are in all living bodies, more than three parts of the whole body; a muscle may be reduced to a very small compass, when dried and exhausted of its fluids, so may nerves, which shews that life consists in fluidity and motion; the blood then, in which motion and fluidity are so apparent, and by which all assimulations are carried on, deserves undoubtedly the name of life; the life of the blood is certainly different from the life of a system, but has all the right to the life of a part, and may be called alive as much as a muscle.

The blood shews the property of life, by preserving itself in its fluidity, while confined in the vessels; as when we tie a ligature on both sides of an artery, so that the circulation shall be stopped, yet the blood between will remain fluid for three hours, nay even if we take out the blood vessel, and tie it on both sides, and freeze it, and thaw it again, the whole

will

will be fluid; but if we take the blood
out of the blood veffel, it will coagulate
in a minute or two; being then refifted
by the heterogenous air, which endeavours
to enter between its parts, and which it
will refift by its power of life, by contracting its parts together, as a mufcle does
when pricked, and by that means it exhaufts its vital power, and coagulates, fo
that both thefe phænomena of the blood,
when in and out of the veffels, depend on its
power of life; the one in preferving itfelf,
and the other in refifting powers which
tend to deftroy it.

The coagulation of the blood depending
on its living principle, cannot be looked
upon as repugnant to all chymical and
mechanical proceffes; in like manner,
the contraction of a mufcle cannot be
looked upon as fuch, life being quite different from what we know of other chymical and mechanical powers. That coagulation depends on life, Mr. HUNTER
fhews

shews by the time the blood takes, according to its state, in coagulating, where he found, that the more it is possessed of that principle, the sooner it coagulates. The coagulable lymph, as long as in the vessels, circulates as a fluid, though when out of the vessels, we find it in the form of a solid; when its power of life is gone, in which state it cannot be fit for circulation, it would fill up the blood vessels, if even we could suppose a spontaneous separation from the thinner parts to take place; it is the life which keeps it fluid in the vessels, and the life which coagulates it, when out of the vessels, and which is always accompanied with motion; and it is clear, that parts, when become entirely motionless and solid, are said to be dead; there can be no doubt then, of the blood being alive in the second sense of life.

Though we cannot reason from the circumstances of the blood, or any fluid, after

it

it is taken out of the veſſels, to its quality while in them, (x) yet I may ſay, we are able to determine, even after the blood is taken out of the veſſels, whether it poſſeſſes ſome of the ſtriking properties of life, and we are more juſtified to call it alive, when in the veſſels, after we find it poſſeſs properties of life, even when taken out of them.

If the power of coagulation is proved to be produced by the ſame conſent of parts, as the contraction of a muſcle, after

it

(x) Many chymiſts, I imagine, miſtook, when they give us the real properties of the component parts of the different juices in the animal, by analyſing them, after they are taken out of the body: for the blood and all other juices, as ſoon as they are drawn out of their veſſels, and mixed with the external air and its acids, acquire directly quite other properties, as is evident, from the experiments made on Magneſia by Dr. BLACK and Dr. MAC BRIDE.

it is taken from the body; if we alfo perceive, as above mentioned, that the blood
poffeffes the power of affimulation; if it
alfo poffeffes that power of reacting and
refifting all heterogenous bodies, that threaten to deftroy the body, &c. I fay, then
the blood certainly is intitled to the name
of life, as well as any other part of the
body; it certainly is not poffeffed of all
the properties, which are produced in
fyftems, as a refult of the whole, yet its
exhibiting fome properties of life, entitles
it to be named alive.

Let us now examine the proofs of Mr.
HUNTER upon this fubject, and the objections made to them by Dr. HENDY, *(y)*
for which, we fhall give the reader the
words of Mr. HUNTER, as related by
Dr.

(y) In his Treatife, entitled an Effay on Glandular
Secretion, and an Enquiry into the Opinion of Mr.
HUNTER that the Blood is alive.

Dr. HENDY, and extract Dr. HENDY's objections as concisely as possible.

Mr. HUNTER's first proof is, says Dr. HENDY, " because it unites living parts " in some circumstances, as certainly as " the recent juices of the branch of one " tree, unite with that of another:" to which Dr. HENDY objects, " That the " gardener does not depend on the juices, " but endeavours to oppose certain grafts. " to similar parts of the stocks; which he " applies together, as soon as the incisions " are made, not that he wants the recent " juices, but because the vegetating powers " are still vigorous; and on this account, " he prefers a particular season of the " year."

If life be taken in the sense of growth, (sect. 3.) it is certain, that the fluids or juices of the tree are the very same with that which Dr. HENDY calls the " *vigor of vegetation* ;" for which

E   reason,

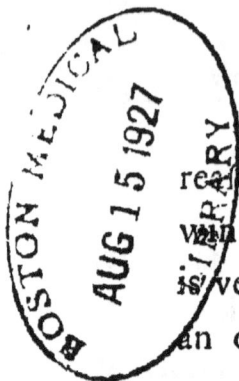

reafon, grafting never can be done in winter, when that life is not in action, is very weak; and for that very reafon too, an old tree, where that juice or life is gone, can never be grafted; Dr. BLAGDEN, in the experiments and obfervations in a heated room, juftly obferves, " that the " fudden melting of fnow which falls " upon grafs, is owing to the vital power " which they poffefs in common with " animals, *(z)* fince that of the adjoining " gravel walk, continue fo many hours " unthawed;" he fays farther, " moift " dead fticks are often found frozen quite " hard, when, in the fame garden, the " tender growing twigs are not at all af- " fected; and many herbaceous vegetables, " of no great fize, refift every winter de- " grees of cold, which are found fufficient " to freeze large bodies of water."

For

(z) And which is their internal motion and heat.

For a further explanation I shall give the reader Mr. HUNTER's words, in a paper read at the Royal Society, June 24, 1775, which are the following :

" To ascertain whether vegetables could " be frozen, and afterwards retain all their " properties when thawed, or had the " same power of generating heat with " animals, I made several experiments. " Vegetable juices when squeezed out of " a green plant, such as cabbage and spin- " nage, froze in a cold about 29°; and " between 29° and 30° thawed again, " which is about 4.° above the point at " which the animal juices freeze and " thaw."

" I. I took a young growing bean, a- " bout three inches long in the stalk, " and put it into the leaden vessel with " common water, and then immersed the " whole into the cold mixture. The wa- " ter very soon froze all round it; how-

E 2 " ever

" ever the bean itfelf took up a longer
" time in freezing than the fame quan-
" tity of water would have done ; yet it
" did freeze, and was afterwards thawed,
" and planted in the ground, but it foon
" withered. The fame experiment was
" made upon the bulbous roots of tulips,
" and with the fame fuccefs."

" II. A young Scotch fir, which had
" two compleat fhoots and a third grow-
" ing, and which confequently was in its
" third year, was put into the cold mix-
" ture which was between 15° and 17°.
" The laft fhoot froze with great diffi-
" culty, which appeared to be owing in
" fome meafure to the repulfion between
" the plant and the water. When thawed,
" the young fhoot was found flaccid. It
" was planted ; the firft and fecond fhoot
" we found retained life, while the third,
" or growing fhoot, withered."

III. " A

" III. A young shoot of growing oats
" with three leaves, had one of the leaves
" put into the cold mixture at 22°,
" and it soon was frozen. The roots were
" next put in, but did not freeze; and
" when put into the ground, the whole
" grew, excepting the leaf which had
" been frozen. The same experiment
" was made upon the leaves and roots of
" a young bean, and attended with the
" same succefs."

" IV. A leaf taken from a growing bean
" was put into the cold mixture, and
" frozen, and afterwards thawed, which
" ferved as a ftandard. Another frefh
" leaf was taken and bent in the middle
" upon itfelf; a fmall fhallow leaden vef-
" fel was put upon the top of the cold
" mixture, and the two leaves put upon
" its bottom; but one half of each leaf
" was not allowed to touch the veffel by
" the bend ; the cold mixture was be-
" tween 17° and 15°, and the atmofphere

22°.

" 22°. The furfaces of the two leaves
" which were in contact with the lead
" were foon frozen in both ; but thofe
" furfaces which rofe at right angles,
" and were therefore only in contact with
" the cold atmofphere, did not freeze in
" equal times; the one that had gone
" through this procefs before, froze much
" fooner than the frefh one. The above
" experiment was repeated when the cold
" mixture was at 25°, 24°, and the at-
" mofphere nearly the fame, and with the
" fame fuccefs ; only the leaves were lon-
" ger in freezing, efpecially the frefh
" leaf."

" V. The vegetable juices above men-
" tioned being frozen in the leaden vef-
" fel, the cold mixture at 28°, and the
" atmofphere the fame, a growing fir-
" fhoot was laid upon the furface, alfo
" a bean-leaf; and upon remaining there
" fome minutes, they were found to have
" thawed

" thawed the furface on which they lay.
" This I thought might arife from the
" greater warmth of thefe fubftances at
" the time of application; but by moving
" the fir-fhoot to another part, we had
" the fame effect produced."

" VI. A frefh leaf of a bean was ex-
" actly weighed ; it was then put into
" the cold atmofphere and frozen. In
" this ftate it was put back into the fame
" fcale, and allowed to thaw. No altera-
" tion in the weight was produced."

" From the foregoing experiments it
" appears; firft, that plants when in a
" ftate of actual vegetation, or even in
" fuch a ftate as to be capable of vege-
" tating under certain circumftances, muft
" be deprived of their principle of vege-
" tation before they can be frozen. Se-
" condly, Vegetables have a power with-
" in themfelves of producing or genera-
" ting heat; but not always in propor-

E 4   " tion

" tion to the diminution of heat by appli-
" cation of cold, fo as to retain at all
" times an uniform degree of heat: for
" the internal temperature of vegetables
" is fufceptible of variations to a much
" greater extent indeed than that of the
" more imperfect animals; but ftill with-
" in certain limits. Beyond thefe limits
" the principle of vegetable, as of animal
" life, refifts any further change. Thirdly,
" the heat of vegetables varies, according
" to the temperature of the medium in
" which they are, which we difcover by
" varying that temperature, and obferv-
" ing the heat of the vegetable. Fourth-
" ly, the expence of the vegetating powers
" in this cafe is proportioned to the ne-
" ceffity, and the whole vegetable powers
" may be exhaufted in this way. Fifthly,
" this power is moft probably in propor-
" tion to the perfection of the plant, the
" natural heat proper to each fpecies, and
" the age of each individual. It may alfo
" perhaps depend, in fome degree, on o-
" ther

" ther circumftances not hitherto obferv-
" ed : for in experiment II. the old fhoot
" did not loofe its powers, while that
" which was young or growing did; and
" in experiment III. and IV. we found,
" that the young growing fhoot of the
" fir was with great difficulty frozen at
" 10°, while a bean-leaf was eafily frozen
" at 22°: and in experiment V. the young
" fhoot of the fir thawed the ice at 28°,
" much fafter than the leaf of the bean.
" Sixthly, it is probably, by means of
" this principle, that vegetables are adapt-
" ed to different climates. Seventhly, that
" fufpenfion of the functions of vegeta-
" ble life, which takes place during the
" winter feafon, is probably owing to their
" being fufceptible of fuch a great vari-
" ation of internal temperature. Eighthly,
" the roots of vegetables are capable of
" refifting cold more than the ftem or
" leaf; therefore, though the ftem be
" killed by cold, the root may be pre- .
" ferved, as daily experience evinces. The

" texture

" texture of vegetables alters very much
" by the lofs of life, efpecially thofe which
" are watery and young; from being bri-
" tle and crifp they become tough and
" flexible. The leaf of a bean when in
" full health is thick and maffy, repels
" water as if greafy, and will often break
" before it is confiderably bent; but if
" it is killed flowly by cold, it will lofe
" all thefe properties, becoming then pli-
" able and flaccid; deprived of its power
" of repelling water, it is eafily made wet,
" and appears like boiled greens. If kil-
" led quickly, by being frozen immedi-
" ately, it will remain in the fame ftate
" as when alive; but upon thawing, will
" immediately lofe all its former texture.
" This is fo remarkable, that it would
" induce one to believe, that it loft confi-
" derably of its fubftance; but from ex-
" periment VI. it is evident that it does
" not. The fame thing happens to a plant
" when

" when killed by electricity *(aa)*. If a
" growing juicy plant receives a ftroke of
" electricity fufficient to kill it, its leaves
" droop, and the whole becomes flexi-
" ble."

All which proves that life and the vegetating powers are one and the fame thing.
I do not know what Dr. HENDY means,
by " oppofing certain grafts to fimilar
" parts," unlefs he thinks that the pores
of the piece grafted and thofe of the tree
may be fo fitted together, that there might
be a free paffage and circulation for the
juices to enter from all fides of the tree to
the piece grafted, fo as to unite it, and
to perform in that piece the action of life,
tho' I do not believe it poffible, to fit the
pores of the piece to thofe of the tree.

Nature

(aa) To kill a whole plant by electricity, it is neceffary to apply the conductor, or give a fhock to every
projecting part ; for any part that is out of the line of
direction will ftill retain life.

Nature grafts, as it were, in the same manner the animal body, by the procefs of granulation and cicatrization, which agree in all circumftances with grafting ; cold and old age are in both cafes unfavourable *(bb)*.

The fecond proof of Mr. HUNTER, is " were either of thefe fluids to be confi- " dered as extraneous or dead matter, " they would act as ftimuli, and no u- " nion would take place, either in the " vegetable or animal kingdom." Here I plainly find that Mr. HUNTER agrees with me in not confining life to the blood a- lone, but in extending it to all fluids in general,

---

(bb) I cannot help making a philofophical remark, namely, that nature always combines, as it were, with every evil a fomething which prevents the very procefs of the evil itfelf. Thus in young perfons where there is danger of general inflammation by the vigour of the blood, that very vigour is the caufe of producing kind fupperation and granulation ; and in old age, when gra- nulation is not favourable, it is commonly the very fame weaknefs which prevents general inflammation.

general, and there is no doubt that all extraneous matter acts as stimuli; the blood itself, as soon as it cannot be as an homogenous principle, constantly changed, and assimulated, (sect. 2.) stimulates the parts it runs to, as is plain from the circumstances of inflammation, where the small vessels are contracted, and constantly stimulated by the flow of blood. In acute rheumatism that kind of stimulation is so strong as to cause (according to my opinion) the constant metastasis.

The testicle of the cock which Mr. Hunter introduced into the belly of a living hen, and found it injected by injecting the liver of the hen, and which he alledges as a proof of his opinion, is objected to by Dr. Hendy, namely, " that since, according to Mr. Hunter's " idea, any dead matter produces a kind " of process, to throw off the dead from " the living part; why did not the tes- " ticle of the cock cause that process to " begin ?

" begin ?" But let me afk Dr. HENDY, why does not the grafted piece deftroy the ftock; the reafon is plain, namely, that the action of the ftock and its juices overcomes the ftimulus of the grafted piece, and gives it life as it were, in uniting it; in the fame manner as the recent tefticle becomes an homogenous part of the hen, and unites with it, inftead of acting upon it as a ftimulus; but that great mafs of fluids in the animal body, were it to be merely paffive and extraneous, and to ferve only to keep the fides of the veffels diftended, it certainly would deftroy the fyftem by its ftimulation, inftead of which, it is now the very principle which guards it from being deftroyed by external ftimuli (fect. 3.).

Mr. HUNTER's third proof is, " that " the blood becomes vafcular, fo that after " amputation the coagulum may be in- " jected by injecting the extremities of " the arteries," which phænomenon Dr.

HENDY

HENDY endeavours to account for, by the elongation of the arteries, and which he thinks is alfo the caufe of union in inflammation, commonly called, *healing by the firft intention*, as we obferve in the adhefions of an inflamed pleura; and in the cafe of DU HAMEL, who placed the fpurs of a cock upon the comb when inflamed, and grew there uniting with the comb; Dr. HENDY therefore fays, " that " it was not the coagulum, but the elon- " gated arteries which run thro' the coa- " gulum, which was injected." I believe all thefe facts will be a ftrong proof to confirm Mr. HUNTER's idea; becaufe it is not enough to know that an elongation takes place (if even we fhould fuppofe it) but we ought to know alfo in what man- ner and by what means it takes place; and is it not to be fuppofed, that elonga- tion is occafioned by the converfion of fluids into folids, which is done by the fluids in general uniting the fides of the veffels, and which Mr. HUNTER pro-

perly

perly calls, the *adhefive inflammation?* Du Hamel's cafe may be explained in the fame mannei. We may farther fay, that the fluids furrounding the fibres are the caufe of that procefs from all its circum-ftances. Inflammations of the extremities do not heal fo kindly, becaufe not furround-ed fo much with living fluids; in young men, when the conftant converfion of folids and fluids is in its greateft vigor, the caufe of growth, kind fuppuration and granulation take oftner place than in old age, and no body ever doubted that old age is combined with a deprivation of the fluids, and youth with a repletion of them; and befides all thefe, I think, in amputation, no elonga-tion of the arteries takes place, and the conclufion of Dr. Hendy, even according to his idea of union in inflammation, I think, is a little too quick, fince that elongation is only in confequence of the furrounding found and living parts, whereas by amputation, when the ftump is expo-fed to extraneous bodies, a contraction,

not

not an elongation, takes place, as we really find, that all the little branches of the arteries in the ftump fhrink, inftead of being elongated; and it is always the nature of living parts, when expofed to extraneous bodies, to contract themfelves, in order to prevent and refift the detrimental effects of the extraneous bodies. It is by the fame law, that the blood coagulates. (fect. 2.)

The fhrinking of the arteries in amputation leads Dr. HUNTER, in his lectures, to blame the method of fecuring each little branch of the arteries in amputation, to prevent an hemorrhage; he fays, it is torturing the patient ufelefsly, becaufe thofe little branches commonly fhrink of their own accord; I believe therefore, that it was the coagula, and not the branches of the arteries, which were injected.

Dr. HENDY goes on further with his objection, " the blood in its coagulated

F   " ftate,

" ftate, cannot poffibly be faid to poffefs
" life." Mr. HUNTER agrees to view
the blood as a fluid, as Mr. HUNTER
expreffes his words, " that in the nature
" of things, there is not a more intimate
" connection between life and folid, than
" between life and fluid, by coagulation
" the blood loofes its fluidity and becomes
" folid; he there fays, that fince life is,
" according to Mr. HUNTER himfelf, that
" property of irratability, by the action
" of any ftimuli, the coagulum has not
" fuch irratability, and therefore cannot
" be alive."

Though it is certain, that *a priori*,
there is no more intimate connection between fluid and life, than between folid
and life, yet I think, *a pofteriori*, we
never find life exift without the union
of the three principles, folid, fluid, and
vapor. (fect. 1.) The blood, in a coagulated ftate, certainly does not poffefs
life, but that principle of coagulation itfelf

felf is owing to its irritability, and is a principle of life, the fame as the irritation and contraction of a mufcle is called its living principle, though when remaining contracted, the mufcle is faid to be dead.

The fourth argument of Mr. HUNTER is very convincing ; his words are, " blood " taken from the arm, in the moft intenfe " cold which the human body can bear, " raifes the thermometer to the fame " height, as blood taken in the moft " fultry heat." It is certain, that the peculiar property of life, to refift both cold and heat, is done by the different circulation of the fluids, (fect. 3.) and therefore cold and heat will not change the degree of warmth in the blood; but were this property to confift in the fibres only, and the blood be merely paffive, then certainly, the moment the blood is drawn out of the fibres, it would be cold or warm, according to the degree of the atmofphere, as all other

dead

dead bodies. The objection of Dr. HENDY, that it cools in lefs than two hours, is I think very futile; Mr. HUNTER never intended to fay, that blood is fo much alive, as never to die; it is fufficient, that after it has been drawn for fome time out of the veffels, and from the animal fyftem, it retains the moft ftriking properties of life, fo as to raife the thermometer to the fame height, in different degrees of heat: Dr. HENDY agrees to call a mufcle alive, when feparated from the animal, as long as it is irritable, and contracts itfelf alternately; though he certainly often faw the mufcle reft in lefs than two hours time, and in my opinion, the life or irritation of the mufcle itfelf confifts in its fluidity, connected with its fibres; a dry mufcle never is irritable. *(cc)*

The

(cc) That is very probable, if we confider that mufcles

The fifth argument is, " blood is ca-
" pable of being acted upon by a ftimulus,
" for it coagulates on expofure, as cer-
" tainly as the cavity of the thorax or
" abdomen inflames from the fame caufe."
Dr. HENDY feems here to laugh at this
argument, as being very ridiculous, and
fays, " we may as well fay a jelly is alive,
" becaufe it coagulates by expofure to air:"
but I think there is a difference between
the coagulation of a jelly, and that of the
blood, becaufe a jelly will coagulate, if
even in a veffel; but the blood, as long
as in the veffels, will refift cold; the coagulation of a jelly is a mere paffive body
to the cold, whereas that of the blood is
done by its living principle of contraction,
as the inflammation of the thorax or ab-

F 3

domen

---

mufcles of animals in full health, abounding with
proper fluids, when cut acrofs, fhew a more remarkable retraction towards each extremity, than the
mufcles of animals in a languifhing ftate and exhaufted
of their fluids.

domen is done by their living principles of contraction and refisting the cold; all jellies will coagulate alike, and in one given time, when in the fame degree of cold, and evaporate to the fame confift-ence; whereas, blood will differ in the time of coagulation, in proportion to the difference of its being found or difeafed, that is, according to its principle of life, as appears from the following argument of Mr. HUNTER, which is the fixth; " The more it is alive, (i. e. the more " the animal is in health) the fooner it " coagulates on expofure, and the more " it has loft of its living principle, as in " cafe of violent inflammation, the lefs it " is fenfible to the ftimulus produced from " its being expofed, and the later it coa-" gulates." Upon which Dr. HENDY re-marks, " It is a very curious kind of life, " that the more it is alive the fooner it " dies, whereas in other kinds of life, we " find the contrary." To which I anfwer, that that very curious and comical kind

of

of life is to be found in all nature, and agrees with the laws of reaction, (fect. 2.) fo that certainly any fubftance dies the fooner, under the fame circumftances, the more it is alive, that is, the ftronger it rea : a thin reed will fuffer the blaft of the moft violent guft of wind, and not be broken; whereas, the great oak tree, by the fame force, will be cruſhed to pieces; nobody ever thought that this is owing to the reed's being ftronger than the oak, but becaufe one yields to the action, and the other reacts; vegetables, and many of the lower clafs of animals, will continue alive in a dormant ftate during the winter without nouriſh-ment, (fect. 3.) whereas the higher clafs of animals will foon die, under the fame circum-ftance; many difeafes will feize a ftrong per-fon, whereas the weak will efcape them, if even expofed to the fame caufes; expofure to cold will be more hurtful to a fanguine, than to a melancholic or hypochondriac temper, fo that weaknefs is not always a

predif-

predifponent caufe to difeafes, or to be affeded by ftimuli, as Dr. HENDY thinks, but on the contrary; Dr. FORDYCE, in his lectures for the fame reafon, makes a difference between irritability and weaknefs; (fect. 2.) after a part has been for a long time difeafed, it certainly loofes that power of reaction, though it is very irritable, fo as to be deftroyed by any ftimuli. It is by this very principle, that the blood, in inflamed cafes, does not react, the ftimulus therefore is not fo great, as that it fhould coagulate, whereas in a found ftate, it reacts with the greateft force, fo as to exhauft its life, and coagulates directly.

Mr. HUNTER does not mean, that the blood never coagulates but by ftimuli; he hardly would commit fuch a grofs miftake, though the infufing of blood in the brain, which Dr. HENDY thought to be a great proof of the blood's coagulating without a ftimulus, might be owing to the ftimuli of

the

the meninges; but death itfelf, in which the blood coagulates, is not done by ftimuli, but by a natural ceffation of life. *(dd)* Mr. HUNTER therefore can only mean, that as long as the blood is alive, it coagulates by ftimuli alfo, and the more

it

(dd) Mr. HUNTER, thinks that even the coagulation in death is produced by a ftimulus; fince he has obferved, that whenever an animal dies fuddenly, as he often kill'd animals by electricity, fo that the living principles are deftroyed at once, that then all the mufcles are in a relaxed ftate, and the blood is not coagulated; whereas when the animal dies a natural death, or in a difeafe, contraction and coagulation take place gradually, by the living powers reacting the ftimulus, namely death; he has made the fame obfervation on a man, who lately died fuddenly in a paffion, that his mufcles were relaxed, and his blood not coagulated; which proves alfo, that the coagulation of the blood, and the contraction of the mufcles, depend on one and the fame principle of life, as I have remarked throughout the whole; but in the experiments on the blood, made by Mr. HEWSON, it feems, that the mere ceffation of motion is able to coagulate the blood. See his experimental Inquiry, &c. page 21.

it is alive, the more it refifts the ftimuli, and the fooner it exhaufts its power of life by this great exertion, and being over-come by the ftimuli it fooner coagulates, not that there are no other means of coa-gulation befides ftimuli.

The feventh argument is very con-vincing; fays Mr. H u n t e r, " the " blood preferves life in different parts " of the body; when the nerves going to " a part are tied or cut, the part be-" comes paralitic, and lofes all power of " motion, but it does not mortify. If " the artery be cut, the part dies, and a " mortification enfues," to which Dr. H e n d y objects, that this will prove no more than that there is nourifhment chiefly conveyed by the arteries, and not by the nerves, and he thinks " that the a-" naftomofing branches of the nerves " continue the circulation of the fluids, " and which depend .upon the action

" of

" of the folids produced by the nervous in-
" fluence." The difference which Dr.
HENDY makes between nourifhment and
life, I have confuted already in the firft
argument, becaufe nourifhment is performed by a conftant converfion of the three
principles in which life confifts, or which
is at leaft never feparable from life ; and
concerning his opinion of anaftomofing
branches of nerves, it may be rejected by
the following arguments :

1ft. Were there any anaftomofing
branches, motion would continue in the
fame manner as the anatomofing branches
of the blood veffels keep up the circulation, fo as to prevent putrefaction, tho'
the large veffels are tied, as is feen by the
application of a *turnicate* upon the bafi-
lick vein, and artery, where the circulation of the hand and wrift is kept up by
the anaftomofing branches.

2dly.

2dly. If the principle nerve is cut off, no anaſtomoſation is to be thought of. (Though it might be ſaid that the nerves which are in the muſcular coat of the blood veſſels, prevent putrefaction.) As ſoon as the communication of the principle nerve is taken from the common ſenſorium or encephalen, all their branches are dead alſo, and have no more communication with the principles of ſenſation.

3dly. There is a very material diffe-rence between nerves and blood veſſels; any ſet of blood veſſels will ſerve in any part of the ſyſtem to keep up cir-culation and reſiſt putrefaction, whereas by nerves we obſerve, that each ſet has its particular office, which another can-not ſupply : the nerves running to the eyes, are fit for viſion, as thoſe to the ear for ſound only, and ſo with the other ſenſes, they have each different feelings;

all

all which muft be accounted for by the different reaction of the different nerves (fect. 2.), according to their nature. Some phyfiologifts carry this point fo far as to attribute certain ftimuli to certain nerves; thus they fay jalap is a fpecific ftimulus to the inteftines; *ipecacuanha*, to the ftomach; *cantharides*, to the neck of the bladder, &c. which opinion if clearly inveftigated, fhews the real difference between the different fets of nerves, fo that anaftomozation cannot be of the fame fervice by nerves, as it is in blood veffels.

The concluding and laft objection of Dr. HENDY is, " that blood cannot per-
" form any action of life when taken
" out of the animal, nor is it organized."
Dr. HENDY calls a mufcle alive as long as it is acted by any ftimuli, though he hardly can mean that a mufcle can perform the action of life, but this objection
tion

tion is entirely owing to his confounding volition with life, which has already been diftinguifhed (fect. 3.). It is true that in fyftems organization and life are infeparable, but were there no life in the component parts of this organization, it could not refult from their mere compofition. See note *(n)*.

S E C T.

SECT. V.

OF PHLEBOTOMY.

IT has been commonly obferved, that the repetition of bleeding produces a plethora, hence a late author would infer, that the frequent ufe of V. S. will occafion all the difeafes, which are the effects of a real plethora, as *convulfions, ft. vitus dance, epilepfy, &c.* Whereas I fhould think, that there is a great difference between a real plethora, and the fudden repletion of the veffels in repeated V. S.

It is true that after bleeding, the procefs of nutrition is very quick, fo that the lofs of blood is foon reftored, which procefs is properly referred to the *vires naturæ medicatrices;* but is it confiftent with
reafon,

reason, that fluids so suddenly secreted, as is in the case of frequent V. S. should be of the same quality and consistence, with fluids secreted in a natural state? A repletion of that kind, might be called a fulness of the vessels with unconcocted fluids, which will produce a set of diseases, quite different from those which are the consequence of a plethora with real blood: a general relaxation will be produced from the rarefaction of the blood, and the preternatural actions of the fibres, in secreting it continually, so that the patient will sink under debility.

If it be true, that the fibres are continually renewed by the blood, (sect. 1.) then their debility must of course increase, when the blood is, by repeated V. S. attenuated; the blood in a rarefied state, certainly does not stimulate the fibres in the same manner, as when in a dense state, by which means the vessels are not excited to contraction, and consequently the

blood

blood is not preffed enough to acquire its proper confiftence.

The rarefaction of the blood, and the weaknefs of the fibres, after repeated *(ee)* V. S. is evident from the ftate of fuch patients who ufe frequent V. S. their appetite is diminifhed inftead of being increafed, and their powers of digeftion are weaken'd.

Mr. HEWSON has plainly fhewn, by his experiments. *(ff)* " that even tempo-
" rary exertion of ftrength in the blood
" veffels, might alter the properties of the
" lymph:" he thought it very difficult to conceive " how the blood veffels fhould

G   " do

(ee) The thinnefs of the coagulable lymph, after repeated V. S. may eafily be found out, by weighing the coagulable lymph, after the different operations, in different cups, where the laft will always be found the lighteft.

(ff) In his Treatife on the Blood, page 129.

" do fo," which may be accounted for
thus; the blood poffeffes life, and re-acts
with this power, to the preffure of the
veffels in circulation, by which it con-
tracts and dilates itfelf alternately, in its
reaction as well as the veffels in their
action, (fect. 2.) and acquires by this
motion its requifite property; but in a
debilitated ftate of the veffels, the reaction
of the blood is ftronger than the preffure
of the veffels, and confequently the blood
begins, by its power of life, to coagulate,
and the lymph is thicker than it is in a
found ftate of the veffels, where it is rarefied by their preffure.

The blood and fluids, which are now
ill-conditioned, will by their connection
with the folids, produce an encreafed
weaknefs, and a diminifhed elafticity in
them, and fo on; fo that the folids and fluids,
in a difeafed ftate, prey upon and deftroy
each other alternately, as they in a found
ftate, nourifh and repair each other.

Again

Again the veffels poffefs a certain power,
with which they contract themfelves continually to a certain point; by a deriva-
tion of blood, the veffels, which muft be
always full, are obliged to contract within
their natural dimenfion; the force then
with which they contract each of their fibres
to prefs upon the blood, is now diminifhed,
great ftrength being wafted by their preternatural contraction, in order to leffen their
diameters, and confequently their preffure
on the blood is not fo great as ufual: hence
the alteration of the blood's property, from
the ftate of the blood veffels.

If the obfervations made by all former
phyficians, that the ftate of the blood is
much altered, according to the nature of
the aliments we make ufe of, be proved
by facts; *(gg)* and yet other experiments
. fhew,

(gg) Which might eafily be proved, by feeding

fhew, (fect. 3.) that all aliments are changed into one and the fame chyle, we then certainly muft account for thofe alterations in the blood, by the ftate of the folids, and by the nature of the aliments and their effects, before they are digefted; to their being eafy or difficult of digeftion. Aliments which require a great exertion of the digeftive powers, will weaken the ftomach, and confequently the folids in general; fo that the fluids will be lefs preffed upon than is requifite, and by that means, be reduced into a vitiated ftate: aliments which are too fufceptible of change, will ferment before they are digefted, and afford little nourifhment to the fyftem; they will alfo weaken the ftomach,

---

fome animals upon fuch food, as is faid to increafe the globulous part, and others upon food, which has been fuppofed to increafe the quantity of the ferum, others again upon food, which is faid to make the juices acrid, &c. and then bleed thefe animals, and analyze their blood, which experiments will be attended with little torture to the animals.

ftomach, by the different eructations they occafion; fo that according to the ftate of the ftomach, in the time or manner of digeftion, and according to the aliments and their effects, the folids will either be weaken'd or ftrengthen'd, and the fluids thereby affected: thus then will the fluids always either be pure or nourifhing, and have their due confiftence, or acrid and vitiated, in proportion to the ftate of the folids.

The fizy or buffy cruft has been always confidered as the criterion of repeating V. S. in inflammatory cafes; but this criterion may often miflead us; becaufe, though it be undeniable, that this appearance is the characteriftic fign of inflammation (the coagulable lymph being rarefied by the heat of the body, and its increafed actions, fwims at the top by its levity) *(bh)* yet whether the cruft indi-

cates

(bh) " The inflammatory cruft or fize is not a new

formed

cates repetition of V.S. till its difappear-
ance, I doubt very much. I have feen
that

---

formed fubftance, but is merely the coagulable lymph,
feparated from the reft of the blood. This feparation
feems to be occafioned by the lymph's being attenuated,
by which means the red particles foon fettle to the
bottom, and leave the furface of the blood tranfparent;
and this tranfparent part being a mixture of the coagulable lymph and the ferum, the former coagulates on its
furface where, in contact with the Air, and the difpo-
fition to coagulate being likewife diminifhed, the blood
remains a long time fluid, and thereby gives time for the
pellicle formed on its furface to attract the reft of the
lymph, and to collect it at the top, leaving the bottom
of the cake proportionable fofter. The fize therefore
is thicker, and denfer, in proportion as the lymph is
thinned, and is lefs difpofed to concrete. But it is not
a certain fign of inflammation, and does not appear to
be the caufe of that difeafe, but only its effect:"
HEWSON's Experimental Enquiry, &c. page 124. I
agree with Mr. HEWSON, in the fact of attenuation of
coagulable lymph; but that very attenuation is a fign
and effect of the increafed action of the veffels, which
is the characteriftic fign of an inflammation. Mr.
HEWSON fays in the fore-mentioned Treatife, page 33.
" as the blood is coagulable by heat, and as the heat of
the

that practice perfifted in, at the expence of the patient's life.

If we confult reafon, we fhall find, that the fizy cruft is not the difeafe itfelf, but its fymptom, pointing out the increafed action and heat of the body, which rare-fies the blood; yet at the fame time, there are many particles of blood ftill in the fyftem, which are not rarefied, (otherwife

G 4   death

---

the animal is increafed in fever, it has been fuppofed, that the blood might be coagulated by the animal heat, even whilft it is circulating in the veffels; but there is little foundation for fuch an opinion, fince the animal heat is naturally only 98° or 100°, and in the moft ardent fever, is not raifed above 100°." Certain it is, that the weaker the blood, the later it coagulates when out of the veffels (fee fect. 4.); but when in the veffels, it may begin to coagulate with a lefs degree of heat, upon quite other principles, I mean, when the veffels prefs upon it preternaturally, as is the cafe in inflammation, it begins then to refift by its power of life, and coagulates infenfibly, though fuch procefs is conti-nually hinder'd, as long as it is in the veffels, by their continual preffure.

death would enfue) and which, in many
cafes, poffefs ftrength enough to refift and
overcome the difeafe, in different ways,
either by converting the very vitiated fluids
into nutritious, or by expelling them
through the different emunctories; but by
repeated V. S. there is every time a great
quantity of found blood drawn out of the
body, by which means we deprive nature
of all its natural ftrength, with which it
ftruggles with the difeafe; the fizy cruft
therefore fhould only ferve to fhew us
the ftate of the inflammation, fo that we
may ufe antiphlogiftics, not that we
fhould repeat bleeding, and exhauft nature of all its ftrength and life. When
bleeding is repeated on the fame criterion
in women with child, we run the rifque
of loofing two lives, that of the mother
and the fœtus (ii); hence I fhould conclude, that whenever bleeding is not neceffary,

---

(ii) Γυναικη ὶν γαςρι' ἰχύση ὑπό τινος των ὀξίων νεσημάτων
Ληφθηα, θανάσιμοι. Hippoc. fect. 5. Aph. xxxi.

ceſſary, as an immediate preſervative of
life, we ſhould uſe it very ſparingly, and
even then, when the caſe does not require
an immediate large evacuation, we ſhould
rather bleed a little, and repeat it, than
draw off a great quantity of blood at
once *(kk)* ; and that we ſhould always obſerve the ſtate of the blood, the
time of its coagulating, the denſity and ,
weight of the lymph ; and laſtly, whether
it abounds with red globules ; becauſe
in the following ſection, I ſhall conjecture,
that many diſeaſes take their origin from
the abundance of that part of the blood,
as others from its want.

S E C T.

(kk) Fere etiam iſta medicina, ubi neceſſaria eſt,
in biduum dividenda eſt. Satius eſt enim, primum
levare ægrum, deinde pepurgare, quam ſimul omni vi
effuſa fortaſſe præcipitare. Celſ. de Med. lib. 2.
chap. 10. See alſo Hip. Aph. ſect. 1. xxiii. and
Celſ. Aph. Sect. 10, 23, to 29.

S E C T. VI.

Of the Red Globules.

THAT the red globules of the blood abound with more oily fubftances, than the ferum and lymph, is evident, from their being more inflamable than the lymph and ferum, and from their yielding large proportions of empyreumatic oil, when analyzed; but this will not prove, that oily particles are the only fubftances out of which the red globules confift, nay it feems more probable, that they are a mixture of oily and earthy particles, by which means their fpecific gravity is more than that of the ferum and lymph; the earthy may ferve as a nucleus to the oily particles, round which they adhere, and which gives them their

globulous

globulous form: the earth of the bones, which abound alſo with oil, may probably take its origin from the red globules, as the fibrous part of the animal, may take its origin from the lymph and ſerum (ſect. 1.)

It is well known, that the beginning of oſſification is produced, by ſome parts of the blood ruſhing continually between the fibres of the fœtus, called cartilages and membranes, and depoſiting their boney matter there in different lamina, which is evident from the inflammation in thoſe parts where oſſification appears *(ll)* ; it is therefore

_____

(*ll*) Anatomiſts do not exprefs themſelves precifely, when they ſay, " that blood veſſels run into the bones," ſince ſtrictly ſpeaking, the bones run between the blood veſſels; the fibres in general being prior in formation to the bones, and the latter are formed between the former; namely, by the calcarious earth which is in the blood, and which earth is continually propelled by the motion of all the fibres together, ſo as to confine

fore very probable, that the red globules
are the very particles which shoot between
the fibres, and form bones; that they
are a mixture of mucilage or oil, and
calcarious earth, some have more muci-
laginous substance than calcarious earth,
and others more earthy than oily parts; the
former constitute the soft, and the latter
the hard bones; where the latter shoot,
they

confine it between them, and form bones; and it is
very probable, that the different convolutions of the
fibres, which run between the bones, give the bones
their laminated and reticular, or net-like form. In the
same manner, we may perhaps be mistaken, when we
say, " that the notches of some bones arise from the
motion and friction of the muscles upon them;" whereas it may be rather said, that the muscles being prior
in appearance to bones, therefore the bony particles
are lodged in their very first formation, by the direction
and situation of the muscles, in such a manner as to
assume the appearance of notches; because, had the
bone, in its original formation been without the notch,
the rubbing of the muscle upon it, would have rather
served to destroy itself, than to alter the figure of the
bone.

they diftend by their hardnefs the fibres, between which they run, and give them the appearance of membranes; whereas thofe fibres between which the former fhoot, remain in their cartilaginous ftate.

Dr. Nesbit's doctrine of the two kinds of offification, between cartilages and membranes is generally accepted, without affigning a reafon, why thofe, which fhoot between membranes, fhould be hard, and others, which fhoot between cartilages, foft bones; whereas, if we pervert the pofition, as I do, and make Dr. Nesbit's caufe the effect, we may rationally account for the difference of membranes and cartilages, fince it feems, that originally thefe membranes were cartilages. Certain it is, that many membranes are in the fœtus originally, as the *dura* and *pia mater*, the membranes of the eyes, the *pleura, mediaftinum, pericardium*, the external coats of the glands, and all mucous membranes, &c. but that

may

may not be the cafe of thofe membranes, between which the bones fhoot, as the contrary fuppofition is very rational.

I further fuppofe, that offification goes on continually, even after growth, though infenfible, and that always fome new boney particles are formed by calcarious earth of the blood; whilft others are wafted and thrown off, infenfibly, in the fame manner, as many phyfiologifts juftly maintain, that the fibrous parts of the animal undergoes a continual granulation.

Even in the univerfe, I believe, that the minerals, fimilar to bones in animals, undergo the fame continual change, though infenfible (fect. 1.), as well as the foft parts, earth, water and air —— *(mm).*

And

(mm) Mr. GODFREY converted a confiderable quantity of diftilled water into a perfectly dry earth, and
Mr.

And thus I would explain the generation of gravel and calculi, as a fault in offification, fo that the calcarious earth, inftead of being impelled into the bones, is carried along with the chyle or blood, into fome other cavities or glands, and lodged there, as in the kidneys, bladder, ductus, cholidochus, inteftines, &c. If the aliments taken abound with calcarious fubftances, there will be more calcarious earth fecreted and mixed with the blood, than the bones are able to abforb; that earth, when it finds no nucleus, is fometimes expelled along with the urine, in the form of gravel, and in the fame manner it may be expelled from the other glands a-long with their fecretions; whereas, when it finds a nucleus, it forms into a ftone. Many circumftances make this

hypo-

Mr. Woulfe is, by his own experiments, perfectly fatisfied of the fact. Dr. Priestley is able to convert a large quantity of earth into air; fee his Philofophical Empiricifm, page 57.

hypothefis probable; the animal fmell, which calculi exhibit in their calcination, like that of bones; that perfons who are troubled with calculi, are found commonly to have offifications of arteries; that in gouty people, a concretion of earth is found between the bones, commonly call'd chalk; the foftening of bones, which is commonly concomitant with calculous complaints, and different offifications; the nature of calloufes in fractures, and granulation in ulcers *(nn)*; that in old age, different

rent

(nn) The calluffes of bones in fractures, and the nature of anchylofes, feem to be a proof of the conftant granulation even in bones. I cannot imagine, that that kind of offification is defignedly produced in fractures, &c, as a new building, were there not a continual procefs of offification, by the ftated laws of the fyftem; Mr. HUNTER's obfervation on the different ftate of bones in animals, after they have been fed upon madder, is a ftrong evidence of this affertion; fee his Treatife on the Teeth. Nor does it feem probable, that the extravafation of the coagulable lymph in wounds, is occafionally produced as a wife manage-

ment

ferent offifications take place; (becaufe the
boney matter, which otherwife would have
been expelled into the bones, adheres on
the fides of the veffels, when by age the
power of motion is diminifhed). All
thefe circumftances make it very probable,
that calculi are generated by a fault in the
offification, occafioned either by the abun-

H dance

---

ment of the fyftem; but I rather believe, that it is a
neceffary confequence of the continual circulation and
motion of that part of the blood itfelf. The coagulable
lymph, which circulates always as a fluid along with
the other parts of the blood, confequently, when there
is an opening between the fibres, and it is expofed to the
air, coagulates and fticks by its tenacity to their fides,
and combines them together. It feems repugnant to
me to give the fyftem fenfe, or to have the foul continually watching, fo as always to produce fomething
new, whenever an accident happens; which is, only
in other words, reviving the old doctrine of STAHL,
who attributed all thefe merits to the foul, which,
as he thought, is continually watching to fuftain and
repair the body. Whereas we may account for all
thefe phænomena, by the ftated mechanical and chymical laws of the fyftem, affifted by life.

dance of the calcarious earth, or by the weaknefs of the fibres, to impel it into the bones. It feems, that the excruciating pains in the gout are occafioned by the little particles of earth, which lodge between the mufcular fibres, and ftimulate them, and which particles, in a found ftate, would have been impelled into the bone: gravelly calculi and gouty complaints are therefore concomitant difeafes, fince they arife from one caufe, namely, a fault in the conftant granulation of bones; hence every thing which weakens or relaxes the folids, as intenfe thought, a fedentary life, the exertion of paffions, intoxication, &c. are the forerunners of all the above-mentioned diforders *(oo).*

The

(oo) Sometimes by a too great fulnefs the veffels will be weakened, and the fame diforders will enfue, fince hæmorrhides, the gravel, &c. are often concomitant complaints; and in which cafes, they commonly aggravate one another, by the connection the rectum has with the bladder, fo that fometimes real

ftrangury

The gout will be hereditary, becaufe the fibres are weak as it were, in their firft make, I mean in the femen of the father.

When once a fecretion has taken place in fome particular part or cavity of the body, all homogenous parts will flow to that place, according to what we obferve in nature in general, and which is very evident in the animal body, from iffues, &c. hence, if fome boney particles have found their way into fome cavity, an influx of the fame will continually take place, and the calculi will be increafed.

That calculi do not affume the internal figure of bones, can be no objection to their being of the fame nature; becaufe they do not fhoot between fibres as bones

do,

---

ftrangury is occafioned by the internal hæmorrhides, when the patient attributes it to its gravelly complaint: an eminent phyfician affured me of the fact, after I told him my thoughts upon the fubject *a priori*.

do, which fibres confine them to certain directions as before-mentioned (note *ll)*; but calculi being left, as it were, to themselves, they adhere together, according to the laws of attraction or impulse, in almoft a round figure, unlefs they have fome heterogenous body to their bafis, as hair, &c. when they take on the form of their bones by adhering round it.

If thefe conjectures could be proved by facts, then certainly it would be of the greateft confequence, to pay particular attention to the nature and quantity of the red globules in V. S. in order to prevent thofe difeafes, which may arife from either their want *(pp)* or abundance.

(pp) Dr. HUNTER juftly fuppofes that the rickets are owing to a want of the animal earth.

F   I   N   I   S.